Building Chic

Excerpted from
A Storey's Guide to Raising Chickens,
by Gail Damerow

CONTENTS

Introduction

Chicken-coop designs are as varied as the people who keep chickens. The best design for you depends on how many chickens you keep, your purpose in keeping them, their breed, your geographic location, and how much money you want to spend. A handy way to get ideas is to find successful chicken keepers in your area — or correspond with those who live in a similar climate — and pick their brains as to what works and what doesn't.

Twelve Tips
for a Successful Coop Design

Some people provide their flocks with perfectly adequate housing by converting unused toolsheds, doghouses, or camper shells. Others go all out, such as the fellow I knew in California who built a two-story structure, complete with a cupola, for his fancy bantams.

No matter how it's designed, though, a successful coop:

- is easy to clean
- has good drainage
- protects the flock from wind and sun
- keeps out rodents, wild birds, and predatory animals
- provides adequate space for the flock size
- is well ventilated
- is free of drafts
- maintains a uniform temperature
- has a place where birds can roost
- has nests that entice hens to lay indoors
- offers plenty of light — natural and artificial
- includes sanitary feed and water stations

Ensuring Good Drainage

If your soil is neither sandy nor gravelly, locate your coop at the top of a slight hill or on a slope, where puddles won't collect when it rains. A south-facing slope, open to full sunlight, dries fastest after a rain. Capture that light and warmth from the sun inside the coop with windows on the south side.

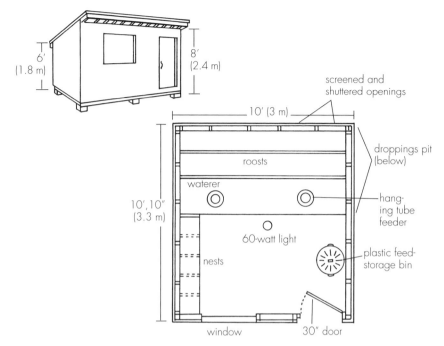

This basic coop plan features roosts over a droppings pit for good sanitation, a window for light, and screened and shuttered openings on the north side to control ventilation. To expand the interior floor space, build the nests on the outside of the coop (see page 12).

Planning for Easy Access

Simple, open housing is easier to clean than a coop with numerous nooks and crannies. If your coop is tall enough for you to stand in, you'll be inclined to clean it as often as necessary. If you prefer a low coop (for economic reasons or to retain your flock's body heat in a cold climate), design the coop like a chest freezer, with a hinged roof you can open for cleaning.

The coop should have both a chicken-sized door and a people-sized door. The chicken door can be a 10-inch-wide by 13-inch-high (25 x 32.5 cm) flap cut into a side wall, opening downward to form a ramp for birds to use when they enter and exit. To keep predators out, the door should have a secure latch that you can fasten shut in the evening, after your chickens have gone to roost.

Space Requirements

The more room your chickens have, the healthier and more content they'll be. Except in extremely cold climates, home flocks are rarely housed entirely indoors but have room to roam outside whenever they please. Yet even in the best of climates, chickens may sometimes prefer to remain indoors due to rain, extreme cold, or extreme heat.

Minimum space requirements, including those shown in the chart below, indicate the least amount of indoor space birds need when they can't or won't go outside for an extended period of time. Birds that never have access to an outside run will do better if you give them more space than the absolute minimum. On the other hand, birds that spend most of their time outdoors, coming in only at night to roost, will do nicely with less space.

To encourage chickens to spend most of their daytime hours outdoors, even in poor weather, give them a covered area adjoining the coop where they can loll out of rain, wind, and sun. Encouraging your chickens to stay out in the fresh air has two advantages: They will be healthier, and their coop will stay cleaner.

Minimum Space Requirements

| Birds | Age | Open Housing | |
		sq ft/Bird	Birds/sq m
Heavy	1 day to 1 week	—	—
	1–8 weeks	1.0	10
	9–15* weeks	2.0	5
	21 weeks and up	4.0	3
Light	1 day to 1 week	—	—
	1–11 weeks	1.0	10
	12–20 weeks	2.0	5
	21 weeks and up	3.0	3
Bantam	1 day to 1 week	—	—
	1–11 weeks	0.6	15
	12–20 weeks	1.5	7
	21 weeks and up	2.0	5

*or age of slaughter

From *The Chicken Health Handbook,* by Gail Damerow (see page 32)

Providing Adequate Ventilation

The more time chickens spend indoors, the more important ventilation becomes. Ventilation serves six essential functions:

- supplies oxygen-laden fresh air
- removes heat released during breathing
- removes moisture from the air (released during breathing or evaporated from droppings)
- removes harmful gases (carbon dioxide released during breathing or ammonia evaporated from droppings)
- removes dust particles suspended in the air
- dilutes disease-causing organisms in the air

Ventilation Quick Check

Use your nose and eyes to check for proper ventilation. If you smell ammonia fumes and see thick cobwebs, your coop is not adequately ventilated.

Confined Housing		Cages	
sq ft/Bird	Birds/sq m	sq in/Bird	sq cm/Bird
0.5	20	(Do not house heavy breeds on wire.)	
2.5	4		
5.0	2		
10.0	1		
0.5	20	25	160
2.5	4	45	290
5.0	2	60	390
7.5	1.5	75	480
0.3	30	20	130
1.5	7	40	260
3.5	3	55	360
5.0	2	70	450

Compared to other animals, chickens have a high respiration rate, which causes them to use up available oxygen quickly while at the same time releasing large amounts of carbon dioxide, heat, and moisture. As a result, chickens are highly susceptible to respiratory problems. Stale air inside the henhouse makes a bad situation worse — airborne disease-carrying microorganisms become concentrated more quickly in stale air than in fresh air.

Ventilation holes near the ceiling along the south and north walls give warm, moist air a way to escape. Screens over the holes will keep out wild birds, which may carry parasites or disease. Dropdown covers, hinged at the bottom and latched at the top, let you close off ventilation holes as needed.

Cold-Weather Concerns

During cold weather, you not only have to provide good ventilation, but you also have to worry about drafts. Close the ventilation holes on the north side, keeping the holes on the south side open except when the weather turns bitter cold.

Warm-Weather Concerns

Cross ventilation is needed in warm weather to keep birds cool and to remove moisture. The warmer the air becomes, the more moisture it can hold. In the summer, leave all the ventilation holes open and open windows on the north and south walls. Windows should be covered with ¾-inch (1.9 cm) screen to keep out wild birds and should slide or tilt so they can be opened easily. Provide at least 1 square foot of window for each 10 square feet of floor space (or 1 sq m of window per 10 sq m of floor space).

Where temperatures soar during summer, you may need a fan to further improve ventilation. Henhouse fans come in two styles: ceiling mounted and wall mounted. The former needs be no more than an inexpensive variable-speed Casablanca (paddle) fan to keep the air moving. A paddle fan benefits birds only if ventilation holes are open and will keep hot air from getting trapped against the ceiling.

A wall-mounted fan sucks stale air out, causing fresh air to be drawn in. The fan, rated in cubic feet per minute, or cfm, should move 5 cubic feet (0.15 cu m) of air per minute *per bird.* If your flock is housed on litter, place the fan outlet near the floor, where it will readily suck out dust as well as stale air. Since some dust will stick to the fan itself, a wall-mounted fan needs frequent cleaning with a vacuum and/or pressure air hose.

Temperature Control

A chicken's body operates most efficiently at temperatures between 70° and 75°F (21–24°C). For each degree of increase, broilers eat 1 percent less, which causes a drop in average weight gain. Egg production may rise slightly, but eggs become smaller and have thinner shells. When the temperature exceeds 95°F (35°C), birds may die.

To keep the coop from getting too hot, treat the roof and walls with insulation, such as 1½-inch (3.8 cm) styrofoam sheets, particularly on the south and west sides. Cover the insulation with plywood or other material your chickens can't pick to pieces. To reflect heat, use aluminum roofing or light-colored composite roofing and paint the outside of the coop white. Plant trees or install awnings to shade the building. An awning can also provide a shady place for birds to rest.

To enhance heat retention in winter, build the north side of your coop into a hill or stack bales of straw against the north wall. Where cold weather is neither intense nor prolonged, double-walled construction that provides dead-air spaces may be adequate to retain the heat generated by your flock. In colder weather, you'll need insulation and, to keep moisture from collecting and dripping, a continuous vapor barrier along the walls. Windows on the south wall will supply solar heat on sunny days (but should be shaded in hot weather).

Cold-Weather Warnings

In winter, rapidly disappearing feed may signal that your chickens are too cold. Eliminate indoor drafts and increase the carbohydrates in the scratch mix. However, it may also mean that the chickens are infested with worms — take a sample of droppings to your veterinarian for testing. Of course, disappearing feed may not be your chickens' fault at all — make sure that rodents, opossums, wild birds, and other creatures are not dipping their snouts into the trough.

Choosing the Right Flooring

Henhouse flooring can be one of four basic kinds: dirt, wood, droppings boards, or concrete. Which you choose will depend on your budget, the siting of your coop, and how much time you're willing to invest in keeping the floor clean.

Dirt

A dirt floor is the cheapest and easiest to "install," but consider it only if you have sandy soil to ensure adequate drainage. Dirt draws heat away, which can be a benefit in warm weather but a potentially dangerous drawback in cold weather. A coop with a dirt floor is not easy to clean and cannot be made rodent proof.

Wood

A floor built from wood planking offers an economical way to protect birds from rodents as long as the floor is at least 1 foot (30 cm) off the ground to discourage mice and rats from taking up residence in the space beneath it. However, wood floors are hard to clean, especially because the cracks between the boards invariably become packed with filth.

Droppings Boards

Droppings boards of sturdy welded wire or closely spaced wooden battens allow droppings to fall through to the bottom of the coop, where chickens can't pick in them. If you opt for droppings boards, not only will the chickens remain healthier, but droppings will be easier to remove because they won't get trampled and packed down. Start with a wooden framework and to it fasten either welded wire or 1 x 2 lumber, placed on edge for rigidity with 1-inch (2.5 cm) gaps between boards. Build manageable sections you can easily remove so that you can take them outdoors and clean them with a high-pressure air or water spray and dry them in direct sunlight. Like wood flooring, droppings boards must be high enough off the ground to discourage rodents.

Concrete

Finished concrete is the most expensive option for a floor, but it's also the most impervious to rodents and the easiest to clean. As a low-cost alternative, mix one part cement with three parts rock-free (or sifted) dirt and spread 4 to 6 inches (10–15 cm) over plain dirt. Level the mixed soil and use a dirt tamper to pound it smooth. Mist the floor lightly with water and let it set for several days. You'll end up with a firm floor that's easy to maintain.

Putting Down Bedding

Bedding, scattered over the floor or under droppings boards, offers numerous advantages: It absorbs moisture and droppings, it cushions the birds' feet, and it controls temperature by insulating the birds from the ground.

Good bedding, or litter, has these properties:

- is inexpensive
- is durable
- is lightweight
- is absorbent
- dries quickly
- is easy to handle
- doesn't pack readily
- has medium-sized particles
- is low in thermal conductivity
- is free of mustiness and mold
- has not been treated with toxic chemicals
- makes good compost and fertilizer

Of all the different kinds of litter I've tried over the years, wood shavings (especially pine) remain my favorite because they're inexpensive and easy to manage. Straw must be chopped; otherwise, it mats easily and, when it combines with manure, creates an impenetrable mass. Of the kinds of straw, wheat is the best, followed by rye, oat, and buckwheat, in that order. Any of these, chopped and mixed with shredded corncobs and stalks, makes nice loose, fluffy bedding.

Rice hulls and peanut hulls are cheap in some areas, but neither material is absorbent enough to make good litter. Dried leaves are sometimes plentiful, but they pack too readily to make good bedding. If you have access to lots of newsprint and a shredder, you've got the makings of inexpensive bedding that's at least as good as rice or peanut hulls, although it tends to mat and to retain moisture; in some areas, shredded paper is sold by the bale.

If you don't use droppings boards, start young birds on bedding a minimum of 4 inches (10 cm) deep and work up to 8 inches (20 cm) by the time the birds are mature. Deep litter insulates chickens in the winter and lets them burrow in to keep cool in the summer.

When litter around the doorway, under the roosts, or around feeders becomes packed, break it up with a hoe or rake. Around waterers or doorways, remove wet patches of litter and add fresh, dry litter (and fix the leak, if any, that caused the litter to become saturated).

If you use droppings boards, after each cleaning spread at least 2 inches (5 cm) of litter beneath the boards to absorb moisture from droppings. An easy-to-manage combination is to place droppings boards beneath perches where the majority of droppings accumulate and to have open litter everywhere else. Your chickens won't be able to get to the manure piles beneath the droppings boards, but they can dust and scratch in the open-litter area, stirring up the bedding and keeping it light and loose.

Earning Their Keep

Scratch — a feed mixture containing at least two kinds of grain, one of them usually cracked corn — can be used to trick chickens into stirring up their coop's bedding to keep it loose and dry. Toss a handful over the litter once a day (traditionally, late in the afternoon when the birds are thinking of going to roost), and your chickens will scramble for it.

Installing Roosts

Wild chickens roost in trees. Many of our domestic breeds are too heavy to fly up into a tree, but they like to perch off the ground nevertheless. You can make a perch from an old ladder or anything else strong enough to hold chickens and rough enough for them to grip without being so splintery as to injure their feet. If you use new lumber, round off the corners so your chickens can wrap their toes around it. Plastic pipe and metal pipe do not make good roosts;

they're too smooth for chickens to grasp firmly. Besides, given a choice, chickens prefer to roost on something flat, like a 2 x 4.

The perch should be about 2 inches (5 cm) across for regular-sized chickens, or no less than 1 inch (2.5 cm) across for bantams. Allow 8 to 9 inches (20 cm) of perching space for each chicken, 10 inches (25 cm) if you raise one of the larger breeds. If one perch doesn't offer enough roosting space, install additional roosts. Place them 2 feet (60 cm) above the floor and at least 18 inches (45 cm) from the nearest parallel wall, spacing them 18 inches (45 cm) apart. If floor space is limited, step-stair roosts 12 inches (30 cm) apart vertically and horizontally, so chickens can easily hop from lower to higher rungs. Either way, make perches removable for easy cleanup and place droppings boards beneath them.

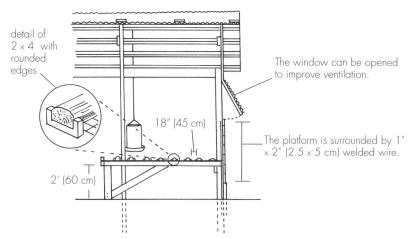

detail of 2 x 4 with rounded edges

The window can be opened to improve ventilation.

18″ (45 cm)

The platform is surrounded by 1″ x 2″ (2.5 x 5 cm) welded wire.

2′ (60 cm)

This roost is made from 2 x 4 lumber with rounded edges, mounted for easy cleaning and spaced 18″ (45 cm) apart, over a raised platform surrounded by wire mesh to keep chickens from picking in their droppings.

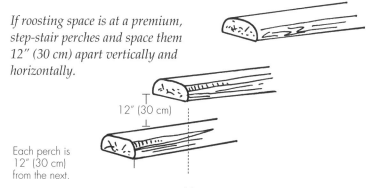

If roosting space is at a premium, step-stair perches and space them 12″ (30 cm) apart vertically and horizontally.

12″ (30 cm)

Each perch is 12″ (30 cm) from the next.

Nests and Nest Boxes

Hens, by nature, like to lay their eggs in dark, out-of-the-way places. Nest boxes encourage hens to lay eggs where you can find them and where the eggs will stay clean and unbroken. Furnish one nest for every four hens in your flock. A good size for Leghorn-type layers is 12 inches wide by 14 inches high by 12 inches deep (30 x 35 x 30 cm). For heavier breeds, make nests 14 inches wide by 14 inches high by 12 inches deep (35 x 35 x 30 cm); for bantams, 10 inches wide by 12 inches high by 10 inches deep (25 x 30 x 25 cm).

A perch just below the entrance to the nest gives hens a place to land before entering, helping keep the nests clean. A 4-inch (10 cm) sill along the bottom edge of each nest prevents eggs from rolling out and holds in nesting material. Pad each nest with soft, clean litter and change it often.

Place nests on the ground until your pullets get accustomed to using them, then firmly attach the nests 18 to 20 inches (45–50 cm) off the ground. Raising nests discourages chickens from scratching in them and possibly dirtying or breaking eggs. Further discourage non-laying activity by placing nests on the darkest wall of your coop. Construct a 45-degree sloped roof above nests to keep birds from roosting on top. Better yet, build nests to jut outside the coop and provide access from the back — chickens won't be able to roost over nests, they'll have more floor space, and you'll be able to collect eggs without disturbing your flock.

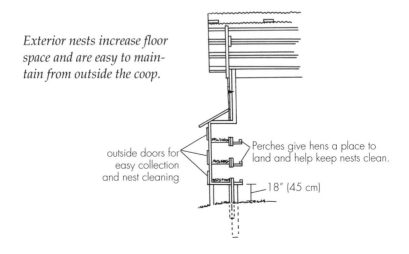

Exterior nests increase floor space and are easy to maintain from outside the coop.

outside doors for easy collection and nest cleaning

Perches give hens a place to land and help keep nests clean.

18" (45 cm)

An alternative plan for creating darkened nests that are easy to clean is to place a long, bottomless nest box on a shelf. Partition the inside of the box into a series of nesting cubicles, with their entrances facing the wall. Allow an 8-inch (20 cm) gap between the wall and the entrances so hens can walk along the shelf at the back. Build a sloped roof above the shelf to prevent roosting. Add a drop panel at the front of the box for egg collection. To clean the nests, make sure no eggs or hens are inside, then pull the box off the shelf and the nesting material will fall out.

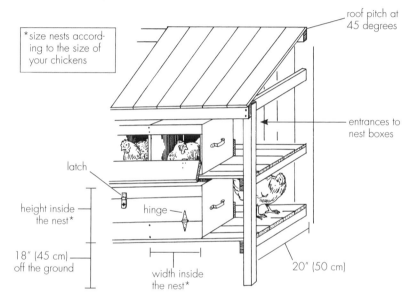

Nest boxes on shelves provide darkened entrances at the back and can easily be cleaned by sliding each box off its shelf.

Choosing and Siting a Feeder

Feeders come in many different styles — the two most common are a long trough and a hanging tube. Regardless of its design, a good feeder has these important features:

- discourages billing out
- prevents contamination with droppings
- is easy to clean
- doesn't allow feed to get wet

Chickens are notorious feed wasters. Feeders that encourage wastage are narrow or shallow and/or lack a lip that prevents chickens from billing out — using their beaks to scoop feed onto the ground. A feeder with a rolled or bent-in edge reduces billing out. To further discourage billing out, raise the feeder to the height of the chickens' backs. The best way to keep a feeder at the right height as a flock grows is to hang it from the ceiling by chains.

A good feeder discourages chickens from roosting on top and contaminating feed with droppings. A trough mounted on a wall allows little room for roosting. A free-standing trough may be fitted with an anti-roosting device that turns and dumps a chicken trying to perch on it. A tube feeder should be fitted with a sloped cover to prevent roosting; unfortunately, most tube feeders don't come with covers anymore, but you can fashion one from the lid of a 5-gallon plastic bucket. It may not keep chickens from roosting, but it will keep their droppings out of the feed.

One hanging feeder is enough for up to 30 chickens.

This trough has adjustable-height legs and an anti-roosting reel that rotates and dumps any bird that tries to hop on. Allow 4″ (10 cm) of trough space for each bird, counting both sides if birds can eat from either side.

Replenishing Trough Feeders

If you use a trough feeder, never fill it more than two-thirds full. Chickens waste approximately 30 percent of the feed in a full trough, 10 percent in a two-thirds-full trough, 3 percent in a half-full trough, and approximately 1 percent in a trough that's only one-third full. Obviously, you'll save a lot of money by using more troughs so you can put less feed in each one.

Since you fill a trough from the top and chickens eat from the top, trough feeders tend to collect stale or wet feed at the bottom. Never add fresh feed on top of feed already in the trough. Instead, rake or push the old feed to one side, and empty and scrub the trough at least once a week.

After having used trough feeders for years, I much prefer tube feeders. Since you pour feed into the top and chickens eat from the bottom, feed doesn't sit around getting stale. A tube feeder is fine for pellets or crumbles, but works well for mash only if you fill it no more than two-thirds full. Otherwise the mash may pack and bridge, or remain suspended in the tube, instead of dropping down.

Where Should the Feeder Be Placed?

Placing feeders inside the coop keeps feed from getting wet but encourages chickens to spend too much time indoors. Hanging feeders under a covered outdoor area is ideal for keeping feed out of the rain and for encouraging the flock to spend more time in fresh air. If you have to keep feeders indoors, for good litter management move them every two or three days to prevent concentrated activity in one area.

How Many Feeders Is Enough?

If you feed free choice, put out enough feeders so at least one-third of the flock can eat at the same time. If you feed on a restricted basis, you'll need enough feeders so the whole flock can eat at once.

Choosing and Siting a Water Source

A chicken drinks often throughout the day, sipping a little each time. A chicken's body contains more than 50 percent water, and an egg is 65 percent water. A bird, therefore, needs access to fresh drinking water at all times in order for its body to function properly. A hen that is deprived of water for 24 hours may take another 24 hours to recover. A hen deprived of water for 36 hours may go into a molt followed by a long period of poor laying from which she may never recover.

Depending on the weather and on the bird's size, each chicken drinks between 1 and 2 cups (237–474 ml) of water each day — layers drink twice as much as nonlayers. In warm weather, a chicken may drink two to four times more than usual. When a flock's water needs go up during warm weather and the water supply remains the same, water deprivation can result. Water deprivation can also occur in winter if the water supply freezes.

Even when there's plenty of water, chickens can be deprived if they simply don't like the taste. Medications, for example, can cause chickens not to drink. Do not medicate water when chickens are under high stress, such as during hot weather or during a show.

Large amounts of dissolved minerals can also make water taste unpleasant to chickens. If you suspect your water supply contains a high concentration of minerals, have the water tested. If total dissolved solids exceed 1,000 parts per million (ppm), look for an alternative source of water for your flock.

Temperature Controls

Chickens prefer water at temperatures between 50° and 55°F (10–13°C). The warmer the water, the less they'll drink. In summer, put out extra waterers and keep them in the shade, and/or bring your flock fresh, cool water often. In cold weather, make sure water does not freeze: Bring your flock warm water at least twice a day (but avoid increasing humidity in the coop by filling indoor fountains with steaming water), use an immersion heater in water troughs, place metal fountains on pan heaters, and wrap heating coils around automatic watering pipes. Water-warming devices are available through farm stores and livestock-supply catalogs.

To keep water from freezing indoors, set a metal fount on a thermostatically controlled heating pan like this one.

Outdoors, drop a sinking heater, like this one, into the water bucket.

Choosing the Right Waterer

Chickens should not have to get their drinking water from puddles or other stagnant, unhealthful sources but should be given fresh, clean water in suitable containers. Provide enough of these waterers so at least one-third of your birds can drink at the same time. No less than once a week, clean and disinfect waterers with a solution of chlorine bleach.

Waterers, like feeders, come in many different styles. The best drinkers have these features in common:

- They hold enough to water a flock for an entire day.
- They keep water clean and free of droppings.
- They don't leak or drip.

Automatic, or piped-in, water is the best kind because it never runs out. But piped-in water isn't without disadvantages. Aside from the expense of running plumbing to the chicken coop, water pipes can leak if not properly installed and freeze in winter unless buried below the frost line or wrapped in electrical heating tape. Automatic drinkers can also become clogged, and so must be checked at least daily.

Properly managed piped-in water is handy for birds kept in cages, since it saves the trouble of having to distribute water to each cage by hand. Automatic devices come in two basic designs: nipples and cups.

Nipples dispense water when manipulated by an individual bird. Since birds have to learn how to drink from a nipple, you'll need to spend time watching to make sure all your birds know how to drink and helping those having trouble. One nipple serves up to five birds.

Cups hold a small amount of water, the level of which is controlled by a valve that releases water each time a bird drinks. Cups come in small and large sizes. The smaller size is for birds in cages. The larger size can be used by a flock. Provide one large cup for up to 100 birds.

Inexpensive plastic 1-gallon (3.7 l) drinkers are fine for young birds, but they don't hold enough for many older birds (and usually get knocked over by rambunctious adults). In addition, plastic cracks after a time, and the cost of constantly replacing those inexpensive waterers adds up fast.

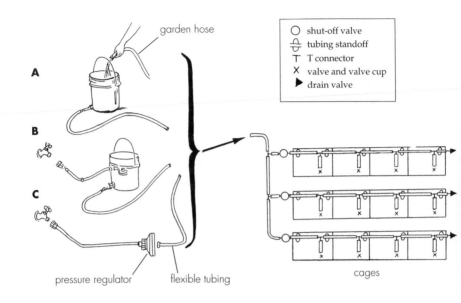

A — garden hose

B

C — pressure regulator / flexible tubing

cages

○ shut-off valve
⊕ tubing standoff
T T connector
X valve and valve cup
▶ drain valve

Water enters automatic waterers in one of three ways: through (A) manually filled tank, (B) float valve-regulated tank, or (C) pressure-regulated direct water supply.

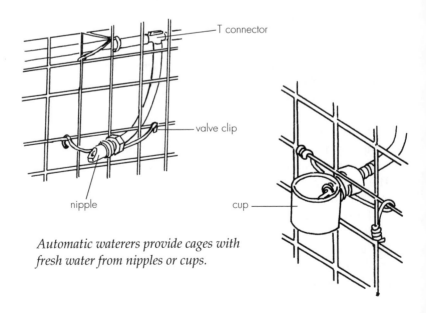

T connector

valve clip

nipple

cup

Automatic waterers provide cages with fresh water from nipples or cups.

Metal waterers are sturdier than plastic and come in larger sizes holding 3 gallons (11 l), 5 gallons (19 l), or more. As in all things, you get what you pay for — a cheap metal waterer will rust through faster than a quality drinker. Whether you use plastic or metal, set the container on a level surface so the water won't drip out. To help birds drink and to reduce litter contamination, set the top edge of the waterer at approximately the height of the birds' backs.

Where Should the Waterer Be Placed?

Placing the container over a droppings pit confines spills so chickens can't walk or peck in moist, unhealthful soil. Build a wooden frame of ½" x 12" x 42" (12.5 x 304 x 1066 mm) boards. Staple strong wire mesh to one side and set the box, wire side up, on a bed of sand or gravel. Place the waterer on top so chickens have to hop up onto the mesh to get a drink.

Homemade Waterer

Make an inexpensive 1-gallon (3.7 l) metal waterer from an empty number 10 can, available for the asking from many cafeterias and restaurants. For the base, you'll need a round cake pan, 4 inches (10 cm) wider in diameter than the can.

Punch or drill two holes in the can, opposite each other and ¾ inch (188 mm) from the open end. Fill the can with water, cover it with the upside-down cake pan, and flip the whole thing over. The little holes let water dribble out every time a chicken takes a sip, keeping the pan filled with fresh water.

¾" (188 mm)

Tips and Advice for
Keeping Chickens in Cages

Chickens may be housed in cages for any number of reasons. Commercial laying hens are caged so their diet can be controlled and so they will be protected from disease, predators, and the weather; layer cages have sloping wire floors so eggs will roll to the outside, where they remain clean and easy to collect. Cocks may be caged to keep them from fighting with one another and/or harassing hens. Exhibition birds are caged for some of the same reasons, as well as to control breeding and condition birds for the showroom.

Caging birds can often be less expensive than building a coop since the cages can be kept in an existing structure. I used to raise exhibition bantams in cages in our garage, where I could control their diet for peak health and production and where I could be sure their valuable eggs would not be hidden, soiled, or eaten by predators. The birds themselves were protected from predators, provided I closed the garage door at night so vagrant dogs couldn't get under the cages and bite off the birds' feet. To find the cages I wanted, I scoured the classified section of the local newspaper until I located a rabbitry going out of business. I picked up all the cages I needed for a song.

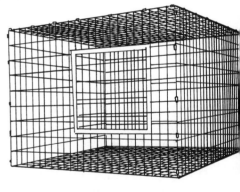

Chickens are often housed in cages so that their diet
behavior can be more strictly controlled; in addition
exhibition birds are often shown in cages.

Making Your Own Cage

Although buying ready-built new cages is sometimes cheaper than making your own — unless you have an inexpensive source for wire — you may find yourself wanting or needing to build a cage for your chickens. Here's how.

Step 1. Gather your materials and tools. You'll need 1-inch by 1-inch (2.5 x 2.5 cm) or 1-inch by 2-inch (2.5 x 5.0 cm) galvanized 12- or 14-gauge welded wire. You'll also need cage clips, or "ferrules;" ferrule-closing pliers; and wire side cutters. If you buy used cages, you may have to redesign them, for which you'll still need the clips, pliers, and cutters.

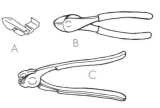

To build or redesign a cage, you will need (A) a latch, (B) wire side cutters, and (C) ferrule-closing pliers, as well as the wire and clips.

Step 2. Cut the wire to the proper dimensions. First, use the chart on page 22 to determine what size cage you need. From welded wire, cut four sides, a top, and a bottom.

Step 3. Assemble the frame. Clip the sides, top, and bottom together along the edges. Around the top and bottom edges, add lengths of 10-gauge wire as reinforcement to keep the cage from sagging.

Step 4. Cut out a door. To make a door, cut a 14-inch-square (35 cm) opening in the center of the front wall, 2 inches (5 cm) from the bottom. File the cut ends smooth. From a separate piece of wire, cut a door 14 inches (35 cm) high and 15 inches (37.5 cm) wide. Using loose ferrules as hinges, attach the door at the bottom, side, or top — the hinge position is strictly a matter of preference, though a door hinged at the bottom so that it drops down when it's opened will leave your hands free. Latches can be fashioned from all sorts of things, but nothing beats the standard cage-door latch in the illustration above.

Step 5. Cover the bottom. Partially cover the cage bottom with a board or piece of heavy cardboard so the birds can rest with their feet off the wire. A resting place is especially important for heavy breeds, since their weight tends to press their feet into the wire. Once a week, scrape droppings off the board or replace the board with a clean one.

Cage Roosts

Lighter breeds appreciate a roost 6 inches (15 cm) off the cage floor. If you do build a roost, you may have to make the cages 6 inches (15 cm) higher so the birds won't rub their combs or top-knots against the ceiling.

Step 6. Secure the cage in place. Cages can be set on sturdy wooden frames or hung from the ceiling in such a way that they won't swing when the birds move around inside. At one time, I made cage legs out of concrete blocks, set on end. Today, I clip cages to the wall with sturdy picture hangers, which allows for easy cleaning beneath them. Removing such a cage to transport a bird, or to put it on the lawn to graze through the wire floor, is equally easy.

Cage Dimensions

Number of Birds	Width	Depth	Height*
1	30 inches (75 cm)	24 inches (60 cm)	24 inches (60 cm)
2	27 inches (68 cm)	32 inches (80 cm)	24 inches (60 cm)
4	46 inches (115 cm)	32 inches (80 cm)	24 inches (60 cm)

*Add 6 inches (15 cm) if you plan to install roosts.

Offering Access to a Yard

A yard offers chickens a safe place to get sunshine and fresh air. Ideally, it should have trees or shrubs for shade, along with some grass or other ground cover. Since chickens invariably decimate the ground cover immediately around their housing, a large yard is better than a small one, but a small one is better than none at all.

Some chicken keepers contend that confining a flock to a coop with properly managed litter and good ventilation is more healthful than letting them into a yard of packed dirt coated with chicken manure. That's certainly true if the hardpan turns to slush in rainy weather. Sad but true, two sure signs of an unsanitary yard are bare spots and mudholes.

Where space for a run is limited, one way to avoid the barren-yard problem is to level the area and cover it with several inches of sand. Go over the sand every day with a grass rake to smooth out dusting holes and remove droppings and other debris. If available yard space is truly minuscule, you might build your chickens a sun-porch with a slat or wire floor and periodically clean away the droppings that accumulate beneath the porch.

On the other hand, if you have plenty of room, keep your flock healthy and take advantage of the cost savings in feed by letting your chickens graze.

Range Rotation

It amazes me that folks who wouldn't dream of planting cabbages or potatoes in the same plot two years in a row never think twice about keeping chickens in the same spot year after year. Even if the coop itself is in constant use — cleaned and disinfected regularly — pathogens and parasites become concentrated in the soil of a constantly used yard.

One way to rotate range is to have two yards that can be accessed from the same doorway, with a gate that can be repositioned so when one yard is opened, the other is blocked off. The disadvantage to this system is that the constant comings and goings of chickens through the single entryway soon kills the grass around the door. When it rains, you have an unsightly and unsanitary situation.

To avoid this mess, you might provide different entries into different yards. By having chicken-sized doors on different sides of the coop, you can periodically close one door and open another. As soon as you switch the chickens to a new yard, rake over their previous entryway, toss on some fresh seed, and let the grass grow while the chickens are away. Rest, sunshine, and plant growth will conspire to sanitize the yard.

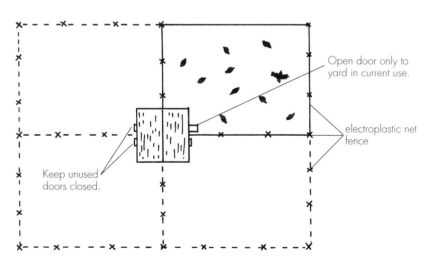

Open door only to yard in current use.

electroplastic net fence

Keep unused doors closed.

To rotate range without moving the housing, put chicken-sized doors on different sides of the coop and separate "ranges" with electroplastic net fencing.

If chickens must constantly use the same entry, a covered door-yard (of the sort used to provide a shady resting place) helps prevent muddy conditions. Alternatively, put a concrete apron in front of the door and periodically scrape off the muck with a flat shovel. As a third alternative, avoid permanent housing altogether and turn your flock out to pasture.

Raising Chickens
Free on the Range

Pasturing chickens on range saves you money by letting your chickens forage for much of their sustenance, and it keeps them healthy by preventing a build-up of parasites and pathogens. But ranging requires a fair amount of ground. It's also labor intensive, since range housing must be moved frequently to new forage areas. Because range housing is moved often, it needs to be light and portable and thus offers little protection against cold weather.

A rudimentary range shelter protects birds from sun and wind. It might be as simple as a roof on posts that can be lifted and moved by two people. For a one-person operation, wheels at one end let you lift the other end and push or pull the shelter to its new location. To protect birds from the elements, enclose those sides of the shelter most subject to prevailing wind and rain.

For our range shelter, we built a plywood corral topped with a lightweight surplus camper shell. Our camper-top shelter is an improvement over open shelters, since it can be closed up at night for predator protection and insulated against cold weather. Although

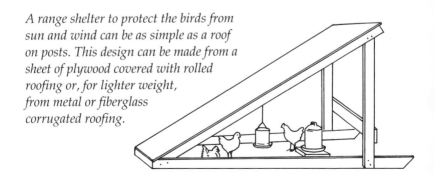

A range shelter to protect the birds from sun and wind can be as simple as a roof on posts. This design can be made from a sheet of plywood covered with rolled roofing or, for lighter weight, from metal or fiberglass corrugated roofing.

we bolted it together so it could be taken apart for storage or moving, we've found it more convenient to move the whole thing in one piece. For this purpose, we added two sturdy hooks on each side. By using straps or chains to connect opposite pairs of hooks, we can lift and move the shelter with our front-end loader. If you don't have a loader handy, a pair of wheels at one end or skids at the bottom of the shelter would help you move it without too much trouble.

Moving a Range Shelter

For total range confinement, move the shelter daily to a new location. For a free-ranged flock, move the shelter when pasture has been grazed down to an inch (2.5 cm) or when bare spots appear.

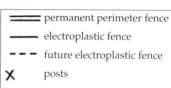

permanent perimeter fence
electroplastic fence
future electroplastic fence
✗ posts

Total range confinement

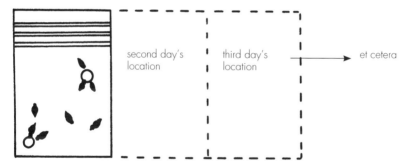

second day's location | third day's location | et cetera

Free range

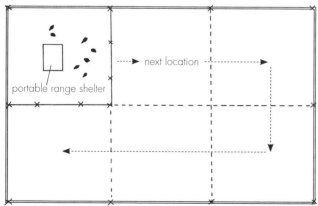

portable range shelter | next location

Range Feeding

Range feeding is one of many old ideas in agriculture that have come around again. Before big business took over the poultry industry, family flocks roamed the back forty, if not the front lawn. Today whole books are written on how to let your chickens eat grass.

As you might expect, short pasture perennials are more suitable for chickens than taller plants. Plants that are in the vegetative, or growing, stage are more nutritious than tough, stemmy plants. Unless your flock follows some other kind of livestock in grazing rotation, during times of rapid vegetative growth — when plants grow faster than the chickens can eat them — you may have to get out the lawn mower to keep the pasture mowed down. Cutting plants short not only keeps them growing but also lets in sunlight, which helps reduce the buildup of infectious organisms.

Chickens tend to stay close to their housing and can quickly overgraze the area, trample the pasture, and destroy plants by digging holes for dust baths. They will be more inclined to forage widely in places where trees give them a sense of security (not to mention shade). Another way to encourage birds to venture forth is to space waterers and feeders at some distance from their housing. You might also scatter scratch grains on the ground, choosing a different place each day so the chickens don't keep scratching up one area. Since foraging causes chickens to burn off extra energy, you can safely feed free-ranged chickens up to ¾ of a pound (360 g) of scratch per two dozen birds.

When the pasture has been grazed down, or if bare spots appear, move the chickens to a new range (by moving their range shelter to new ground or by letting them into a new yard from permanent or semi-portable housing). In doing so, you'll be imitating the natural conditions under which plants evolved and under which they grow best. In nature, a flock moves together to avoid predation, quickly grazes down an area (during which it scratches up the ground and deposits large amounts of manure), then moves on.

How long a flock takes to graze down a given area depends on a number of factors, including the size of the flock, the kind and condition of the pasture, temperature, and rainfall. When sun, rain, and warm weather combine to help plants grow quickly, a flock might graze a given area for two weeks or more. In cool, hot, or dry weather, when plants grow slowly, a flock may graze pasture down to nothing in a matter of days. Chickens that are confined within a range house need to be moved to new ground daily.

How small an area you can confine your flock to, and how long you can keep them there, can be determined only through watchful experimentation within these limits:

- Let chickens in when plants are no more than 5 inches (12.5 cm) tall.
- Move chickens out when plants have been grazed down to 1 inch (2.5 cm) or when bare spots appear.
- The stocking rate, or number of chickens one acre can support in a season, for a free-ranged flock on well-managed pasture is 200 birds per acre (500 per hectare).
- The stocking rate for a confined flock on well-managed pasture is 500 birds per acre (1,250 per hectare).
- The longer a range shelter stays in one place, the more time it takes for the pasture to be restored once the shelter is moved.

Over the years, pasture soil will increase in acidity. When soil pH drops below 5.5 as determined by a soil test, spread lime at the rate of 2 tons per acre (4.5 metric tons per hectare). Then let the pasture rest to give plants time to rejuvenate and to break the cycle of parasitic worms and infectious diseases.

To optimize feed-cost savings, move a range shelter as soon as you notice ration consumption starting to rise.

Helping Your Chickens Find Their Shelter

Chickens can be decidedly stupid about finding their home after it has been moved. You can help them along by watching for stragglers that insist on bedding down in the old place and by never moving the house far outside the previous range — chickens are conservative by nature, and don't like to venture more than about 200 yards (180 m) from their home place.

To avoid the problem of having chickens forget where they live, you might construct the range shelter so that the flock is entirely confined by it. An advantage to range confinement is its superior ability to exclude predators. Its disadvantage is the need to move the shelter more often. Since the unit gives chickens less space to roam, you don't have to move it as far — only to the nearest patch of fresh pasture.

Introducing Chickens to the Range

You can put young birds on range as soon as they feather out, raising them away from older birds while they develop immunities through gradual exposure to the disease in their environment. In a warm climate, a flock can be kept on range year-round. Where winter weather turns nasty, the flock may need to be moved to permanent housing during the colder months.

Despite its many advantages, a distinct disadvantage to putting chickens out to pasture is their greater susceptibility to predators. A good fence goes a long way toward solving the predator problem.

Fencing

Whether your chickens have only a small yard or are free to roam the range, you'll need a stout fence to keep them from showing up where they aren't wanted and to protect them from predators. The fence should be at least 4 feet (120 cm) high so predators won't climb over and chickens won't fly out. It may need to be higher if you raise flyers such as Leghorns, Hamburgs, Old English, or many of the bantams.

The ideal chicken fence is made from tightly strung, small-mesh woven wire. The best fence I ever had was a 5-foot-high (150 cm) chain-link fence that once came with a house I moved into. A chain-link fence isn't one I can recommend, though, because of the exorbitant cost of building a new one.

The most common coop fencing material is "chicken wire," or "poultry netting," which consists of 1-inch (2.5 cm) mesh woven in a honeycomb pattern. Take care to specify galvanized wire designed for *outdoor* use. Chicken wire designed for indoor use rusts away all too soon when used for outdoor fencing. Many people use so-called "turkey wire," which has 2-inch (50 mm) mesh, because it's cheaper than chicken wire, but it doesn't hold its shape as well.

A better though more expensive option is yard-and-garden fencing, which has 1-inch (2.5 cm) spaces at the bottom that graduate to wider spaces toward the top (thus using less wire to keep the cost down). The smaller openings at the bottom keep small chickens from slipping out and small predators from slipping in. Birds and predators can't sneak under if you pull the fence tight and attach it to firm posts that don't wobble. As further insurance, place pressure-treated boards along the ground and staple them to the bottom of the fence.

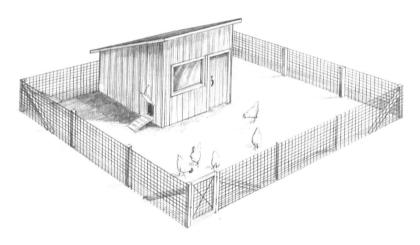

The ideal chicken fence is made from tightly strung, small-mesh woven wire. Yard-and-garden fencing, with 1-inch (2.5 cm) spaces at the bottom that graduate to wider spaces toward the top (thus using less wire to keep the cost down), makes a good secure fence for the coop. The smaller openings at the bottom keep small chickens from slipping out and small predators from slipping in.

Downy chicks can pop right through most fences, but they won't stray far if they have a mother hen inside the yard to call them back. Chicks outside the fence are, however, vulnerable to passing dogs and cats. To keep them in, get a roll of 12-inch-wide (30 cm) aviary netting, which looks just like chicken wire but has openings half the size. Attach the aviary netting securely along the bottom of your regular woven-wire fence.

Electric Fencing

If you need additional protection from dogs and other predators, string electrified scare wires along the top and outside bottom, 8 inches (20 cm) away from your fence. The top wire keeps critters from climbing over, and the bottom wire discourages them from snooping along the fence, pushing against it, or attempting to dig under it.

If you're putting up electric wires anyway, consider building an all-electric fence. We have used electric fencing to successfully confine our chickens for many years. It's relatively inexpensive and virtually predator proof. "Virtually" doesn't mean "entirely" — even the best fence won't stop hungry hawks or opossums.

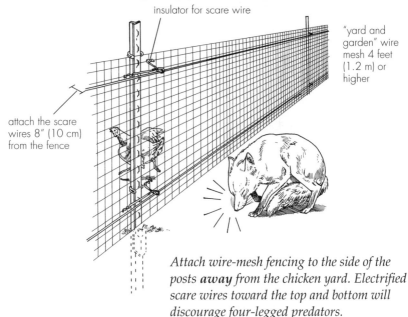

insulator for scare wire

"yard and garden" wire mesh 4 feet (1.2 m) or higher

attach the scare wires 8" (10 cm) from the fence

*Attach wire-mesh fencing to the side of the posts **away** from the chicken yard. Electrified scare wires toward the top and bottom will discourage four-legged predators.*

One good electrified chicken fence is made from electroplastic netting. Chickens can see the netting more easily than they can see individual horizontal electric wires, and the fence comes completely preassembled so it's easy to move when the flock needs fresh ground. Electroplastic net for poultry comes in two basic heights: the shorter version for sedate breeds is a little over 20 inches (50 cm) high; the taller version is 42 inches (105 cm) high.

A controller (the device transmitting electrical energy to the fence) that plugs in will give the fence more zap, especially when fast-growing weeds drain its power, but out in a field you can use a battery-operated energizer of the sort sold to control grazing livestock or to protect gardens. A lightweight net fence with a battery-powered controller lets you easily move the fence each time you move the shelter.

Although chickens aren't as susceptible to getting zapped as other livestock because of their small feet and protective feathers, they do learn to respect an electric fence. But first they have to know their home territory. Whenever you move chickens to a new coop, confine them inside for at least a day. When you let them into the yard, they won't stray far from home.

Other Storey Titles You Will Enjoy

Chicken Coops, by Judy Pangman.
A collection of hen hideaways to spark your
imagination and inspire you to begin building.
180 pages. Paper. ISBN 978-1-58017-627-9.
Hardcover. ISBN 978-1-58017-631-6.

The Chicken Health Handbook,
by Gail Damerow.
A must-have reference to help the small flock
owner identify, treat, and prevent diseases common
to chickens of all ages and sizes.
352 pages. Paper. ISBN 978-0-88266-611-2.

Storey's Guide to Raising Chickens,
by Gail Damerow.
A wealth of expert information about every stage
of chicken ownership, from starting a backyard flock
to putting eggs on the table.
448 pages. Paper. ISBN 978-1-60342-469-1.

Storey's Illustrated Guide to Poultry Breeds,
by Carol Ekarius.
A definitive presentation of more than 120 barnyard
fowl, complete with full-color photographs
and detailed descriptions.
288 pages. Paper. ISBN 978-1-58017-667-5.
Hardcover with jacket. ISBN 978-1-58017-668-1.

**Your Chickens: A Kid's Guide to Raising
and Showing,** by Gail Damerow.
Friendly and encouraging advice for young chicken
owners everywhere – for ages 9 and up.
160 pages. Paper. ISBN 978-0-88266-823-9.

These and other books from Storey Publishing are available
wherever quality books are sold or by calling 1-800-441-5700.
Visit us at *www.storey.com*.